"The two most important days in your life are the day you are born and the day you find out why"

Mark Twain

In the Greater Scheme of Things

Whispers of the Soul

Revised Edition

J. Carlos Aguirre

authorHOUSE®

AuthorHouse™
1663 Liberty Drive
Bloomington, IN 47403
www.authorhouse.com
Phone: 1 (800) 839-8640

Published by AuthorHouse 06/11/2018

ISBN: 978-1-4389-6446-1 (sc)

Print information available on the last page.

This book is printed on acid-free paper.

Dedication

To Maharishi Mahesh Yogi, the teacher who led me to the portals of my own mind and a greater universe.

Preface

Hymn to Dionysus

> How they toss about in the immense universe as they whirl and seek each other, these myriad souls which burst forth from the great soul of the World! They fall from planet to planet, and in the abyss mourn the forgotten homeland... They are your tears, Dionysus... O Great Spirit, O Divine Liberator, take your daughters back into your heart of light.
>
> -Orphic fragment

Ancient hymns echo a lost memory of our origins, of a timelessness that spans worlds. We need only to listen to that deep, inner silence, to the whispers of our soul, to discover our own ancient secret that has been proclaimed down through the ages in verse and song. If we are to awaken from our amnesia, we must look deep within ourselves, beyond the din of our daily mental chatter, beyond our biological identities. We are not just biological machines confined to limited boundaries only as some might suggest. We need only to seek in the laboratory of our own being to discover that amidst all the chaos of the world there is a point of infinity that stretches omni-directionally to all times and places. We are those "tears of that Great Spirit" that have forgotten our homeland—Eternity / Infinity!

Prologue

Philosopher as "Cosmic Scientist"

The traditional view of the philosopher in western culture is of the thinker, spectator and dialectician or "mental gymnast." He or she is relegated to the halls of academia among the dusty shelves of outworn books with little relevance for the everyday person struggling to survive and make sense of a seemingly chaotic world.

The new view proposed in this book is one of the philosopher as investigator into consciousness, into the very fabric of the universe; tracing the footprints of creation from within the Self, which is the Self of the universe. He or she is an amalgam of the scientist and the artist in probing and appreciating nature's deeper secrets. They are the "transcendentalists" of the world whose vision pierces all parochialisms and dogmatisms of thought.

The Cosmic Scientist as a pioneer of consciousness systematically follows the reverse flow of the mind to the source of thought, to the origin of Being.

The philosopher's role is to imbibe the qualities of the "Source" or "Unified Field" that is, "the Good and the Beautiful" and radiate those qualities in his or her day to day activities. Life is to be lived not thought about. There is an old saying, "Knowledge in the books, stays in the books," in other words, theories or information by themselves are of little use if they are not our experience.

The 'lover of wisdom" is the lover of the truth of life and what is the truth of life? – *That* – which is unchanging beyond the realm of change yet found at the heart of all change. THIS is what Plato and other Pre-Socratic philosophers experienced, not just thought about! And this same experience is available to everyone through a system of deep meditation that transcends all thought.

Introduction

In the Greater Scheme of Things... is an affirmation of the grand or cosmic order and the interconnectedness of all life. It's so easy to get caught up in the day-to-day details of earning a living and to miss the 'bigger picture' of the fantastic miracle we call life. We often miss the forest for the trees. If we aim at mere survival, we become pessimists; should we expand our vision beyond our micro-concerns, we become dreamers; but should we integrate those visions with our daily life, we become whole and successful as human beings.

This book is a celebration of the new renaissance of the human spirit in a time when science, technology, and insights as ancient as mankind are beginning to merge on a common ground or "transcendental territory"—a meeting point of science and metaphysics.

Scientific advancement in quantum physics has shown the limitations of classical, linear thinking in grasping the grand order at the subatomic level, and the images from the Hubble telescope have helped to push the boundaries of human imagination on the macro or cosmic scale.

One of the promises of a new, emerging twenty-first-century humanity is growth beyond a more dogmatic mindset through the blending of a scientific and spiritual understanding in approaching that glibly stated term, 'reality.'

This understanding would be more akin to Albert Einstein's notion of the "cosmic religious sense." Einstein noted that this cosmic religious sense "which recognizes neither dogmas nor God made in man's image" has been a distinguishing characteristic of some of the great geniuses of the past, whose expanded vision enabled them to see a more sublime relation between science and religion.

Or as one modern sage aptly stated, "**Those who think that that religion is outside of science, neither knows science nor religion.**"

There is so much talk about current high-tech advancements, yet without a doubt, the most awesome instrument in the known universe is the human nervous system.The mere fact that the number of interconnections between all brain cells is about 10 x 10^{800} (10 followed by eight hundred zeros) compared with 10^{100} (10 followed by one hundred zeros) estimated atoms in the known universe gives one pause to ponder. As one scientist has noted, a single brain cell may be likened to a personal computer. Our brains have anywhere between 10 billion to 100 billion neurons or brain cells, which would make our brains an ultra-computer by comparison.

So why don't we see and understand more? Consider the fact that our sensory range within the electromagnetic spectrum is primarily limited to the visible light portion of the spectrum, a rather small sliver of the grand pie. We are practically oblivious to 98% of the entire range, from the longer wavelengths of radio / infrared to the shorter wavelengths of gamma / cosmic rays. Couple this with the fact that our brains take in 400 billion bits of information per second, but we are only aware of 2000 bits of information. Of all the incoming sensory signals, less than .000000005% reaches our conscious minds, or in the words of William James (American philosopher and psychologist),

"Compared with what we ought to be, we are only half awake... We are making use of only a small part of our possible mental and physical resources."

This gives us a hint of what potential lies within all of us. No wonder, as Einstein stated, that a creative person with a cosmic religious sense **"feels the individual destiny as an imprisonment and seeks to experience the totality of existence as a unity full of significance."**

Keep in mind, do you live of your own power? A basic course in human anatomy and physiology reveals a marvel of delicate structure and balance that is the beyond the province of our analytical mind to fully comprehend. We are upheld by an invisible matrix of energy in which we move and have our being. Life is truly a mystery and wonder that transcends our anthropocentric field of vision; as Einstein also noted, "**The most beautiful thing we can experience is the mysterious. It is the source of all true art and science.**"

The following "affirmations" and quotes should be read prefaced with the phrase "In the Greater Scheme of Things..." in mind as a way of elevating one's perspective from the conditioning of our everyday bias toward overwhelming materialism. In this way, it is the hope of the author to inspire the reader and expand his or her perspective, if even momentarily, to a higher order of 'the way of things'—a celebration of that grand matrix of existence that we call life or being.

When reading the following quotes, some may resonate with our sense of how things are, while others may cause some dissonance with our present understanding or mindset. Either case can be a means of slipping past the mind-matter dualism of thought-bound consciousness, if only by our "remembering," through the 'silent whispers of the soul,' that innermost wellspring that is the basis of intuition.

Consider the fact that our opinions are usually limited by our culturally rooted past conditioning and that:

> There are more things in heaven and earth, than are dreamt of in your philosophy.
>
> Shakespeare's Hamlet

> Every person takes the limits of their own field of vision for the limits of the world.
>
> Arthur Schopenhauer

In the Greater Scheme of Things...

Contrary to popular opinion, 'dumb luck' is a myth...

> There is no such thing as chance, and what we regard as blind circumstance actually stems from the deepest source of all.
>
> Frederick Von Schiller

> If we are alert, with minds and eyes open, we will see meaning in the commonplace; we will see very real purposes in situations which we might otherwise shrug off and call 'chance.'
>
> Roland Bach

> Every cause has its effect; every effect has its cause; everything happens according to Law; Chance is but a name for the Law not recognized.
>
> The Kybalion

> Fundamentally, nothing in the universe is random.
>
> Dr. J. Haeglin, nuclear physicist

Men have made of luck a phantom—a pretext for their own folly, for rarely does luck conflict with intelligence and most things in life can be set straight by a keen wit and sharp insight.

Democritus

Only shallow people believe in luck, believe in circumstances. It was somebody's name, or he happened to be there at the right time, or it was so then, but another day it would have been otherwise. Strong men believe that things are not by luck, but by law, and there is not a weak or cracked link in the chain that joins the first and the last of things—the cause and the effect.

Emerson

Blind chance is the natural corollary of the modern mindset with its notion of a universe based on randomness and matter as the root cause of existence. Yet even chaos theory reveals an inherent order in nature, and quantum physics demonstrates that all matter is really energy. The deepest examination of existence reveals that all energy is purposeful or guided by intelligence.

By opening our awareness to nature's deepest functioning within ourselves, we discover that synchronistic events unfold in our lives. We discover that nothing in nature is random. I affirm a divine order in the greater scheme of things; "the dice of God are always loaded."

LIFE IS FAIR...

As you sow, so shall you reap.

The Bible

Karma means you don't get away with anything.

Buddha

Law, not confusion, is the dominating principle in the universe, justice, not injustice, is the soul and substance of life.

James Allen

A perfect equity adjusts its balance in all parts of life. The dice of God are always loaded. The world looks like a multiplication-table, or a mathematical equation, which, turn it how you will, balances itself.

Emerson

The results of karma cannot be known by thought, and so should not be speculated about. Thus thinking, one would come to distraction and distress.

Buddha

This is that ancient doctrine of Nemesis, who keeps watch in the universe and lets no offence go unchastised.

Emerson

I cannot believe that the inscrutable universe turns on an axis of suffering; surely the strange beauty of the world must somewhere rest on pure joy.

Louise Bogan

To the materialist point of view, life is conspicuously unfair and without equity. The view that life is unfair is an affirmation of the intellect, a dissection of what is seen, whereas the view that life is inherently fair is an affirmation of the soul, an intuition of the unseen. Life has grander vistas that cannot be properly judged from the perspective of one lifetime; how could it be otherwise if we affirm divinity? Otherwise, we would affirm divinity and injustice in the same breath. I affirm a divine equity in the greater scheme of things...in the larger landscapes of the soul.

CHANGE IS AN INEVITABLE CONSTANT...

Change is our best friend, when we want to unfold the non-changing nature of life.

Maharishi Mahesh Yogi

Everything flows.

Heraclitus

How is it possible to find real happiness by centering ourselves in those things which, by their nature, must pass away? Abiding and real happiness can only be found by centering ourselves in that which is permanent.

James Allen

Change is the handmaiden Nature requires to do her miracles with.

Mark Twain

In blossom today, then scattered, Life is like a delicate flower. How can one expect the fragrance to last forever?

Admiral Ohnishi

Nature is a dynamic process of transformation; nothing in the relative world remains static. All things have a measure of duration or time, a coming into being and decaying—thus the "wheel of birth and death."Change is a wave on the ocean of life, the sea eternal of non-change. When our focus is solely on the world of change, we are lost in the "valley of tears," the world of gain and loss, happiness and sorrow. Life consists of "three worlds": the gross, the subtle, and the transcendental. The rise and fall of civilizations is a measure of the lesser or greater awareness and knowledge of these "three worlds."

WHAT WE CALL THE "I" TRANSCENDS OUR EVERYDAY NOTIONS...

The boundaries of individual life are not restricted to the boundaries of the body, nor even of those of one's family or home; They extend far beyond those spheres to the limitless horizons of unbounded cosmic life.

Maharishi Mahesh Yogi

Man is a stream whose source is hidden. Our being is descending into us from we know not whence.

Emerson

What we commonly call man, the eating, drinking, planting, counting man, misrepresents himself. But the soul whose organ he is, would make our knees bend. When it breathes though his intellect, it is genius; when it breathes though his will, it is virtue; when it flows though his affections, it is love.

Emerson

The body is a tomb.

Orphic fragment

The riddle of the self lies in unlocking our identity from our bodily existence only. Our essence is consciousness, and our search for immortality should not be restricted to the body; it should also include the immortality of consciousness as an unbroken continuum—a conscious immortality, beyond the realm of change. This is the area commonly called spirit. I affirm that which 'I am' precedes all materiality.

WE ARE OUR BROTHER'S KEEPER...

The world is my family.

Ancient Sanskrit

We are not independent but interdependent.

Buddha

What do we live for, if not to make life less difficult for each other.

George Eliot

There is a destiny that makes us brothers: None goes his way alone; All that we send into the lives of others comes back into our own.

Edwin Markham

It is one of life's beautiful compensations of this life that no man can sincerely try to help another without helping himself.

Emerson

Mankind has become so much one family that we cannot insure our own prosperity except by insuring that of everybody else.

Bertrand Russell

If fear is the "first-born" of duality, then love is the offspring of unity. I therefore affirm that my brothers / sisters are not just my 'other self,' but are of the greater Self of which I am.

WE SEE THOUGH A GLASS DARKLY...

> You have been told that life is darkness, and in your weariness you echo what has been said by the weary.
>
> K. Gibran

Through the window of our time-bound self, we succumb to the inertia of matter or the contraction-phase of the spirit. True freedom lies in piercing the veil of our ego-consciousness.

◆ SUCCESS' DEMANDS A MORE COMPREHENSIVE DEFINITION ...

> To laugh often and much to win the respect of intelligent people and the affection of children; to earn the appreciation of honest critics and endure the betrayal of false friends; to appreciate beauty, to find the best in others; to leave the world a bit better, whether by a healthy child, a garden patch, or a redeemed social condition; to know that even one life has breathed easier because you have lived. This is to have succeeded.
>
> Ralph Waldo Emerson

> There is in this world no function more important than to be charming—to shed joy around—to shed light upon dark days. Is that not to render service?
>
> Victor Hugo

What is our purpose? What do we live for? These are questions that each of us must answer for ourselves. I affirm that one of the most valued measures of success is peace of mind. How do we attain peace of mind? Live in accordance with our highest nature, in tune with natural law.

AS WE AWAKEN TO A NEW INNER VISION...

To see the world in a grain of sand and heaven in a wildflower: Hold infinity in the palm of your hand and eternity in an hour.

William Blake

If we could see the miracle of a single flower clearly, our whole life would change.

Buddha

And here is my secret, a very simple secret. It is only with the heart that one can see rightly. What is essential is invisible to the eye.

The Little Prince

Until, the breath of this corporeal frame and even the motion of our human blood almost suspended, we are laid asleep in body, and become a living soul; while with an eye made quiet by the power of harmony, and the deep power of joy, we see into the life of things.

Wordsworth ("Tintern Abbey")

By fine-tuning or cultivating our mind/body system, we start to see with 'new eyes.' By contacting the inner silence, we start to see with the eyes of the soul and not just our physical eyes. I affirm that we are rightful inheritors of a more cosmic vision.

LIFE RECYCLES ITSELF...

It is life in quest of life in bodies that fear the grave. There are no graves here. These mountains and plains are but a cradle and a stepping-stone.

K. Gibran

...Death is but the doorway to new life...We live today — we shall live again—In many forms shall we return...

Ancient Egyptian

Death is merely waking from a dream.

Yoga Vashitha

Death is not a period, but a comma.

Anon

Time is but a sea eternal, we drown in it many times and are washed ashore again and again...

H. Rider Haggard

Through the travail of ages, amidst the pomp and toils of war have I fought and strove and perished countless times... As if through a glass darkly, the age old strife I see, For I have fought in many guises, many names, but always me.

General George Patton

Our minds have identified with the elements around us—'what we see, we become.' As a result we are tossed about in the milieu of circumstances. But in the "stream of becoming" there is a hope of renewal of our being, and in renewal or spiritual evolution there lies a bridge to the infinite. I affirm that who I truly am transcends the boundaries of space-time.

OTHING REALLY COMES INTO BEING OR IS UTTERLY DESTROYED...

Fools! They have no far-reaching thoughts who imagine that what was not before can come into being, or that anything can perish and be utterly destroyed.

Empedocles

Wherefore all these things are but the names which mortals have given, believing them to be true.

Parmenides

The unreal has no being; the real never ceases to be. The final truth about them both has thus been perceived by the seers of ultimate Reality.

Vyasa

Energy / Matter can neither be created nor destroyed.

First Law of Thermodynamics

Yet nothing is corrupted or destroyed and quite abolished, but the names trouble men.

The Divine Pymander of Hermes

"Change is an illusion."

Parmenides

The phenomenal world dazzles and befuddles sensory-oriented minds with the incessant becoming and fading of 'things.' To those minds that are aware of the unchanging aspect of life, the changing world is a magician's slight of hand, a moving image of the immovable, a superimposition of the mind, just as thoughts are superimposed on silence or waves onto the ocean. I affirm a world of non-change beyond the sea of phenomena.

WOMANKIND IS NATURE'S CIVILIZING INFLUENCE...

The dignity of a woman's life is infinite—her status immeasurable, her capacity unbounded, her role divine. She fulfills the role of sustainer, creator and promoter of life. She is that focal point from which radiates serenity. In the fast changing world of today, her vision penetrates into the reality of the far beyond, the reality which endures beyond change.

Maharishi Mahesh Yogi

From a sensitive woman's heart springs the happiness of mankind, and from the kindness of her noble spirit comes mankind's affection.

Kahil Gibran

Where women are honored, the gods are pleased. Where they are not honored, all works become fruitless.

Manu

It is through the refinement of emotions that life evolves; it is a highly delicate faculty of experience.

Maharishi Mahesh Yogi

The feminine is the most tender and refined aspect of creation. All creation is a relation of interacting principles. The feminine is the embodiment of reception, refinement, and fecundity. All relative life comes through the agency of the feminine. This is why the ancients honored the personification of life as the Magna Mater or the Great Mother. I affirm and bow at the ever-nourishing feminine at the root of all things.

THE TERM 'PROGRESS' IS SOMEWHAT SUSPECT...

> It has become appallingly obvious that our technology has exceeded our humanity.
>
> Albert Einstein

> The means by which we live have outdistanced the ends for which we live. Our scientific power has outrun our spiritual power. We have guided missiles and misguided men.
>
> Martin Luther King, Jr.,

The intellect holds sway over the present age of man, usurping the more tender qualities of life. The ancients utilized a more subtle instrument of perception, called by some the 'intelligence of the heart,' a more refined mode of consciousness, which could see into the nature of things unimpeded by our quotidian prejudice.

EDUCATION IS MUCH MORE THAN THE EXCHANGE OF FACTS...

(The teacher) If he is indeed wise he does not bid you enter the house of his wisdom, but rather leads you to the threshold of your own mind.

Kahil Gibran

Education has for its object the formation of character.

Herbert Spencer

No technical knowledge can outweigh knowledge of the humanities, in the gaining of which philosophy and history walk hand in hand.

Winston Churchill

We know too much and feel too little. At least, we feel too little of those creative emotions from which a good life springs.

Bertrand Russell

Those who seek truth by means of intellect and learning only get further and further away from it. Not until your thoughts cease all their branching here and there, not till you abandon all thoughts of seeking for something, not till your mind is as

> motionless as wood or stone, will you be on the right road to the Gate.
>
> Huang Po

> Reading makes a full man, meditation a profound man, discourse a clear man.
>
> Benjamin Franklin

> Learn less and contemplate more.
>
> Rene Descartes

Contemporary education has become more training for creature survival than cultivation of our more refined faculties of mind and soul.

The Newtons and the Einsteins of the world realized that the deeper powers of our humanity exceeded the formalisms of society and bordered the unbounded expanses of our soul. To train a person is to show them how to perform a skill. Today's education, on the other hand, should not only impart facts, but should show one how to think critically. Education in the most profound sense should be the cultivation of our entire being or the development of our complete human potential through the unfolding of our pure consciousness, the deepest recesses of our soul.

WE ARE INEXTRICABLY CONNECTED...

We live in succession, in division, in parts, in particles. Meantime within man is the soul of the whole: the wise silence; the universal beauty, to which every part and particle is equally related; the eternal One.

Emerson

He then learns that in going down into the secrets of his own mind he has descended into the secrets of all minds.

Emerson

Listening not to me, but the Logos it is wise to acknowledge that all things are one.

Heraclitus

Realizing the Self (the eternal One) in all and all in the Self, free from egoism and free the sense of 'mine', be happy.

Astavakra Samhita

What many have cognized in the recesses of their being, science is now corroborating in the laboratory experiments of quantum physics, culminating in the most abstract understanding of the unified field. All of life is connected; it is only our vision that is fragmented. This one understanding throws light on seemingly disconnected events and shows that indeed the world is a cosmos or 'beautifully ordered.' I affirm a oneness beyond the veil of our daily perception of things.

LIFE IS AN INFINITE GAME...

Now, life is a very simple game and in the simplicity you enjoy. Life is not for thinking. It is to be lived and enjoyed.

Maharishi

Eternity is a child at play, it is the dominion of a child.

Heraclitus

And we are Pieces of the Game He plays Upon this Chequer-board of Nights and Days; Hither and thither moves, and checks, and wins, And then another Cosmic Game begins.

Omar Khayyam

On the vast canvas of the Self, the picture of the manifold worlds is painted by the Self Itself and that Supreme Self, seeing but Itself, enjoys great delight.

Shankara

There is but one infinite game.

James P. Carse

The man with time is a king and can play whenever and wherever he wants to.

Daniel Blum

The true object of all human life is play. Earth is a task garden; heaven is a playground.

G.K. Chesterton

The man of understanding, the knower of the Self, who plays the game of life has no similarity to the deluded beasts of burden of the world.

Astavakra Samhita

Cosmic play challenges the imagination of the scientist and tickles the heart of a child. To see this play is to awaken to a higher order of being rather than be caught up in the phenomena of living. To try to control things and events is human; to flow with nature is divine. I affirm that my soul plays beyond the boundaries of time and space.

WE ARE NOT OUR EGOS...

We are by nature observers and thereby learners. That is our permanent state.

Emerson

It's going to take a lot of awareness for you to understand that perhaps this thing you call "I" is simply a conglomeration of your past experiences, of your conditioning and programming.

Anthony de Mello

In determining the full potential of man it has been found that whereas the individual personality appears to be bound by time, space and causation, the boundaries of an individual life actually touch the unlimited horizon of eternal life.

Maharishi

A human being is part of the whole, called by us "universe", a part limited in time and space. He experiences himself, his thoughts and feelings, as something separate from the rest—a kind of optical delusion of consciousness. This delusion is a kind of prison for us, restricting us to our personal desires and to affection for a few persons nearest to us.

Our task must be to free ourselves from this prison by widening our circle of compassion to embrace all living creatures and the whole of nature in its beauty.

Albert Einstein

A man is the façade wherein all wisdom and all good abide.

Emerson

Selfish egoism is the root of the wide extending branches of misery in the forest of this world.

Vasistha

Individual identity is fragile in the spiritually immature, as it provides a central focus point for the 'doer' of action. The 'I' is identified with that which is tangible. To the more spiritually mature, action is surrendered to the divine will, and we are but observers of the world of phenomena. I affirm that our true nature is unbounded and infinite consciousness.

WE LIVE IN "PARALLEL UNIVERSES" WITH ONE ANOTHER...

It's not what you look at that matters, it's what you see.

Henry David Thoreau

As a man is, so he sees.

William Blake

A pessimist sees the difficulty in every opportunity; an optimist sees the opportunity in every difficulty.

Sir Winston Churchill

If the only tool you have is a hammer, you tend to see every problem as a nail.

Abraham Maslow

And even as each one of you stands alone in God's knowledge, so must each one of you be alone in his knowledge of God

Kahlil Gibran

Eyes and ears are bad witnesses for those who have barbaric souls.

Heraclitus

'Our world' is a combination of many factors, but none more important than the maturation of our soul or character. We 'see' what we are capable of seeing. The 'soul of the artist' can see the artistic masterpiece of nature, while the 'soul of the non-artist' sees something of less aesthetic value.

WHAT WE BELIEVE TO BE, IS MORE IN OUR HEADS...

> In the last analysis, we see only what we are prepared to see, what we have been taught to see. We eliminate and ignore everything that is not part of our prejudices.
>
> Jean Martin Charcot

> Objectivity has about as much substance as the emperor's new clothes.
>
> Connie Miller

> There are no facts, only interpretations.
>
> **Friedrich Nietzsche**

> Reality is merely an illusion, albeit a very persistent one.
>
> Albert Einstein

We are for the most part unaware of our social 'knowledge filters.' Many more times than not, we reason from a conclusion rather than towards one and in effect self-validate our beliefs. I affirm that it is not until we cease the mental chatter within that can we truly step back from the canvas of our 'self-painted reality' and finally see truly.

TIME IS A CONCEPT TO BE TRANSCENDED...

Time is a moving image of eternity

Plato

Time is a conception to measure eternity.

Maharishi

One who is anciently aware of existence is master of every moment, feels no break since time beyond time in the way life flows.

Lao Tzu

"Today" means boundless and inexhaustible eternity. Periods of months and years and of time in general are ideas of man, who calculate by number; but the true name of eternity is Today.

Philo

A million of our solar years may be less than a minute in the mind of God. We should train ourselves to think in grand terms: Eternity! Infinity!

Paramahansa Yogananda

> The immortal man is he who has detached himself from the things of time by having ascended into that state of consciousness which is fixed and invariable.
>
> James Allen

We are intimately cognizant of the flow of time. But what is time? Time is intimately associated with movement—whether subjective or objective, psychological or 'real.' Only by going beyond all movement within ourselves can we discover the unmovable—the source of all time and our being.

Our task is to become intimate with the timeless aspect of life by transcending the realm of change; that is our joy and goal in life.

LIFE PASSES US BY...

The average man, who does not know what to do with his life, wants another one which will last forever.

Anatole France

Leisure is the mother of philosophy

Thomas Hobbes

But the gods, taking pity on mankind, born to work, laid down the succession of recurring Feasts to restore them from their fatigue, and gave them the Muses, and Apollo as their leader, and Dionysos, as companions in their Feasts, so that nourishing themselves in festive companionship with the gods, they should again stand upright and erect.

Plato

We are caught in the web of minutes, hours, days, living mainly unfulfilled moments; between each of these moments lays a 'gap'—a portal to infinity.

I affirm that it is my birthright to reclaim the infinite expanse within myself; then I become a player with time and not a prisoner of it.

UNITY PERVADES EXISTENCE...

Reality is implicitly whole, it is broken up in our own minds only.

Dr. David Bohm (physicist)

The world of physics is a world of shadows, we were not aware of it; we thought we were dealing with the real world... Subject and object are only one.

Erwin Schrodinger (physicist)

The reason why the world lacks unity, and lies broken and in heaps, is because man is disunited with himself.

Ralph Waldo Emerson

It is the fragmented mind which does not see the thing as it is.

Patanjali

Be in a realm where neither good nor evil exists. Both of them belong to the world of created beings; in the presence of Unity there is neither command nor prohibition.

Abu Yazid al-Bistami

> To those who are awake the world-order is one, common to all, but the sleeping turn aside each into a world of his own.
>
> Heraclitus

There is a unity that is beyond all numbers, yet the source of all numbers. This is unadulterated, pure consciousness, the warp and weft of reality. When we are intimate with this unity within ourselves, then we will know that the apparent multiplicities are but 'emanations' of that unity, like the spokes on a wheel radiating out in all directions.

BELIEFS ARE ONLY ROADMAPS, NOT THE TERRITORY...

A belief is not merely an idea the mind possesses; it is an idea that possesses the mind.

Robert Bolton

Truth is to Belief, as Being is to Becoming.

Plato

Creed is an ossified metaphor.

Elbert Hubbard

So many gods, so many creeds, so many paths that wind and wind, while just the art of being kind, is all the sad world needs.

Ella Wheeler Wilcox

Men never do evil so completely and cheerfully as when they do it from religious conviction.

Blaise Pascal (philosopher and mathematician)

Dogma is an anathema to the soul; it petrifies the creative spirit. Life should become more transparent with increasing vision from our own inner journey, rather than more opaque with the mere re-iteration of others' thinking and handed-down formalisms. I affirm that my soul will not give way to the tyranny of fear and religious demagoguery but will expand into rarified realms of faith pure and knowledge sublime.

TRANSCENDENCE IS THE NATURAL AIM OF MANKIND...

We are quite literally, made to transcend. Inside flap of *The Biology of Transcendence*

By Joseph Chilton Pearce

Expansion of happiness is the purpose of life, and evolution is the process by which it is fulfilled.

Maharishi Mahesh Yogi

We must close our eyes and invoke a new manner of seeing, a wakefulness that is the birthright of us all…

Plotinus

Know thyself and you will know the universe and the gods.

Delphic Oracle

Nothing of the outside world is able to satisfy the mind… such is the glamour of material life. It attracts, but fails to satisfy the thirst for happiness.

Maharishi Mahesh Yogi

Nature will not allow humanity to be deprived of the vision of the transcendental reality for very long.

Maharishi Mahesh Yogi

> It is said that human nature by itself is something small and limited...and how has the infinite been embraced by something so tiny? For not even in our life is man's spiritual nature enclosed within the bounds of the flesh; on the contrary the physical body is limited by its neighboring parts, but the soul expands freely over the whole of creation by means of the activity of thought.
>
> Gregory of Nyssa

> For, as the spirit wakens, it craves more and more to regard all existence not merely with a creature's eyes, but in the universal view, as through the eyes of the creator.
>
> Olaf Stapledon

> Self-transcendence is the law of life: that which ceases to grow, i.e. self-transcend, ceases to be.
>
> E. C. Prophet

Dissatisfaction with the status quo is the natural human condition. The basic urge for that something more varies in concept for each of us. Our thoughts are but a blind grasping in response to this need within us. Our souls seek to awaken from this earthly slumber of death and taxes and expand to become citizens of the universe. The need for transcendence is nothing other than the 'prodigal son' returning home to his divine status.

CONSCIOUSNESS IS PRIMARY...

In the beginning was the Word (Logos or "Organizing Principle") and the Word (Logos or "Organizing Principle") was with God and the Word (Logos or "Organizing Principle") was God

John 1:1

The unmanifest provides the impetus for creation just by becoming aware of itself. This is the start of the entire time-space geometry, the beginning of creation. From consciousness creation begins.

Maharishi Mahesh Yogi

Consciousness is wakefulness, unbounded alertness, pure Intelligence, pure existence, self-referral fullness, all-knowingness-the self-sufficient and unmanifest source, course and goal of creation.

Maharishi Mahesh Yogi

We begin to suspect that the stuff of the world is mind-stuff.

Arthur Eddington (physicist)

...the universe begins to look more like a great thought than a great machine.

Sir James Jeans (astronomer)

> Consciousness is the common denominator underlying the possibility of any philosophy, world view, religious attitude, art, or science. I, therefore, affirm the systematic primacy of consciousness as such.
>
> Franklin Merrell-Wolff (philosopher/mathematician)

> "Get over it and accept the inarguable conclusion. The universe is immaterial-mental and spiritual"
>
> - Richard Conn Henry, Professor of Physics and Astronomy at John Hopkins University

Empirical or materialistic-based science assumes that matter is primary and that consciousness is an epiphenomenon or latecomer in an evolutionary scheme. Current science is symptomatic of a split-personality culture that cannot reconcile material and spiritual worldviews. Ancient sacred science begins where modern science ends—the unified field—that most abstract field which informs all creation. All manifestations are but varying modifications of consciousness or pure, divine intelligence. Consciousness is the first principle of this sacred science and the secret to the marvels of antiquity. Out of the 'well of infinity' springs all the possibilities; this was the true magic of yore.

The world is not what it seems...

The universe is an intellectual concept.

Maharishi Mahesh Yogi

Bell's theorem is a mathematical proof. What it 'proves' is that if the statistical predictions of quantum theory are correct, then some of our commonsense ideas about the world are profoundly mistaken.

Gary Zukav

"Every thing is of the nature of no thing."

Parmenides

For as far as I know, there is no proof whatever of the existence of an objective reality apart from our senses, and I do not see why we should accept the outside world as such solely by virtue of our senses.

M.C. Escher

Nature or what we call the natural world order is but a metaphor for the Absolute.

Author

"If quantum mechanics hasn't profoundly shocked you, you haven't understood it yet. Everything we call real is made of things that cannot be regarded as real."

– Niels Bohr

The world of matter has been shattered by science as it has demonstrated that no material forms are permanent. All matter has 'dissolved' before the discerning eyes of the scientist, and what remains is a sea of fluctuating energy. 'The universe' is constantly changing according to the refinement of the discerning intellect. I affirm a fluid reality, molded in part by the penetrating power of the everyday observer.

WE FASHION 'THE UNIVERSE' IN OUR OWN IMAGE...

The world is as we are.You are to me as I am to myself……I am to you as you are to yourself.

Maharishi Mahesh Yogi

To the dull mind all nature is leaden. To the illumined mind the whole world sparkles with light.

Ralph Waldo Emerson

Each of us literally chooses, by our way of attending to things, what sort of universe we shall appear to ourselves to inhabit.

William James

If a triangle could speak, it would say, that God is eminently triangular, while a circle would say that the divine nature is eminently circular.

Baruch Spinoza

And if oxen and horses and lions had hands and could draw with their hands and do what man can do, horses would draw the gods in the shape of horses, and oxen in the shape of oxen , each giving the gods bodies similar to their own.

Xenophanes

> If God created us in his own image, we have more than reciprocated.
>
> Voltaire

> I cannot imagine a God who rewards and punishes the objects of his creation, whose purposes are modeled after our own -- a God, in short, who is but a reflection of human frailty.
>
> Einstein

Through the faculties of our thoughts and imagination, we bring to bear our image of the world; we 'color' what we see, aided by the cooperative projection of our social milieu. We are micro-creators and destroyers... I affirm that undiscovered worlds await the expanding purview of my soul.

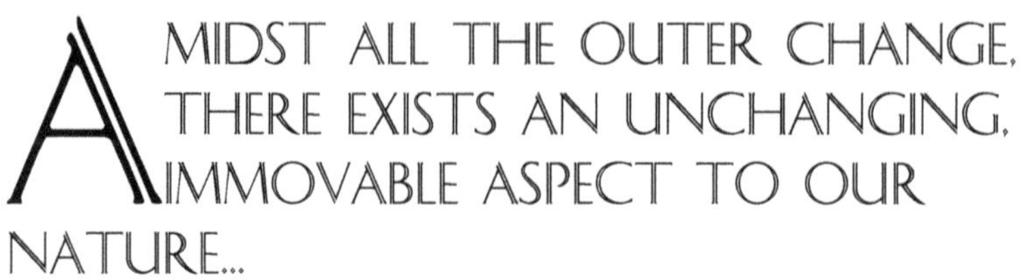
NATURE...

Life is one continuous and homogenous whole. The waves of individual life on the ocean of life arise without breaking the continuity and all pervading status of eternal, absolute Being.

Maharishi Mahesh Yogi

The Self is eternal- without beginning or end. By knowing the Self, you know the secret of immortality.

Katha Upanishad

Give me a place to stand on and I can move the earth.

Archimedes

Everything here, but the soul of man, is a passing shadow. The only enduring substance is within.

William Henry Channing

In the great course of time and human history, knowledge of the absolute aspect of life is lost and rediscovered.This is the knowledge that was whispered in the ancient temples that gave rise to great civilizations. It has been encoded in the myths of the world and sacred architecture.

It is the Great Silence, which is beyond the most subtle and is the source of all wisdom, science, and religion.

LIFE IS BUT A WAKING DREAM...

Our normal waking consciousness, rational consciousness as we call it, is but one special type of consciousness, whilst all about it, parted from it by the filmiest of screens, there lies potential forms of consciousness entirely different.

William James

Some day comes the Great Awakening when we realize that this life is no more than a dream. Yet the foolish go on thinking they are awake: Surveying the panorama of life with such a clarity, they call this one a prince and that one a peasant What a delusion! The great Confucius and you are both a dream. And I, who say that all this is a dream, I, too, am a dream.

Chuang Tzu

The immortal man is as one who has awakened out of a dream. He is a man with knowledge, the knowledge of both states—that of persistence, and that of immortality—and is in full possession of himself.

James Allen

Dreams are real as long as they last. Can we say more of life?

Henry Havelock Ellis

Continuity of experience and a common vocabulary are the lynchpins of defining our waking world. But denizens sharing the common boundaries of 'the cave' will proclaim that their subterranean dwelling defines the limits of the world. Enlightenment is seeing beyond those limits—outside 'the cave' of our conditioned existence.

THE WORLD IS FULL OF OPINION...

Human opinions are toys for children.

Heraclitus

Do not seek truth, simply cease to cherish opinion.

Zen Saying

Those who know do not speak, those who speak, do not know.

Lao Tzu

Loyalty to a petrified opinion never yet broke a chain or freed a human soul.

Mark Twain

To men, some things are good and some are bad. But to God, all things are good and beautiful and just.

Heraclitus

Wisdom means keeping a sense of the fallibility of all our views and opinions, and the uncertainty and instability of the things we most count on.

Gerald Brenan

Opinions arise out of the world of shadows, images cast before the judging ego. The soul does not judge but observes the world of change as a silent witness. It is only from the deep well of silence that true judgment can be rendered with impartiality. I affirm a world beyond human opinion, beyond good and evil.

WE ARE STRANGERS IN A STRANGE LAND...

Strange is our situation upon earth. Each of us comes for a short visit, not knowing why, yet sometimes seeming to a divine purpose.

Albert Einstein

Man is a fallen god who remembers the heavens.

Alphonse Delamartin

I am a child of earth and starry heaven, but my race is of heaven alone.

Orphic fragment

The first man is out of the earth and made of dust; the second man is out of heaven.

I Corinthians 15: 47-49

That an earthly man is a mortal god and that the heavenly god is an immortal man.

The Divine Pymander of Hermes

I, too am... a fugitive from the gods and a wanderer—because I put my trust in raging Strife.

Empedocles

How they toss about in the immense universe as they whirl and seek each other, these myriad

> souls which burst forth from the great soul of the World! They fall from planet to planet, and in the abyss mourn the forgotten homeland... They are your tears, Dionysos... O Great Spirit, O Divine Liberator, take your daughters back into your heart of light.
>
> Hymn to Dionysos

Deep within our psyche lays the answer to our origins, an answer that defies any textbook notions. We are ever-ancient and ever-young, beyond the boundaries of time and space. Listen to the silent whispers of your soul and realize that we are timeless. It has been said in the most ancient traditions of the world that we are the sons and daughters of immortality. "Learning is but a remembering..."

OUR DAILY LIVES MAY BE LIKENED TO A HYPNOTIC STATE...

> From the womb to the tomb our minds are molded by standardized templates for behavior, emotions, attitudes and ultimately thought. Our beliefs, values, social status, personality, and often even individual interests are programmed right into us almost without our noticing.
>
> Stuart Litvak and Wayne A. Senzee

> To those who are awake the world-order is one, common to all; but the sleeping turn aside each into a world of his own.
>
> Heraclitus

> Those who awake live in a state of constant amazement.
>
> Buddha

Our overly sensory-oriented culture has displaced us from our inner selves.

The din of our daily lives drowns out the silent wisdom of our souls. We are more akin to sleepwalkers than to ones alert to 'the way of things.' The world mesmerizes us so that we continue to be caught up with the baubles and wares of the 'world bazaar.'

THE BOUNDARIES OF ANIMATE LIFE EXPAND WITH DEEPENING CONSCIOUSNESS...

> Stars are best regarded as living organisms, but organisms which are physiologically and psychologically of a very peculiar kind.
>
> Olaf Stapledon

> The sun...is probably inhabited...by beings whose organs are adapted to the peculiar circumstances of that vast globe.
>
> William Herschel (discoverer of the planet Uranus)

> When each of the stars necessary for the constitution of Time had obtained a motion adapted to its condition, and their bodies bound or encompassed by living chains, had become beings possessing Life, and had learned their prescribed duty. They pursued their course [in space].
>
> Plato, Timaeus

We make distinctions with our categorizing intellect as to what bears 'life' and what does not. With an ever-so-slight shift of perception, we would see a universe much more populated than previously imagined. Consciousness is ubiquitous! I affirm a 'living' and 'breathing' universe beyond our mere mortal notions of the way of things.

LINEAR HISTORY IS A 'MYTH'...

All our history ... is no more than accepted fiction.
Voltaire

It is a superstition of modern thought that the march of knowledge has always been linear.
Sri Aurobindo

What has been will be again, what has been done will be done again, there is nothing new under the sun.
Ecclesiastes 1:9

Poetry comes nearer to vital truth than history.
Plato

The world's great age begins anew, the golden years return, The earth doth like a snake renew Her winter weeds outworn; Heaven smiles, and faiths and empires gleam like wrecks of a dissolving dream.
P.B. Shelly, Hellas

> I existed from all eternity and, behold, I am here; and I shall exist till the end of time, for my being has no end.
>
> Kahlil Gibran, "Anthem of Humanity"

> While probably each art and each science has been developed as far as possible and has again perished, these opinions, with others have been preserved until the present like relics of the ancient treasure.
>
> Aristotle, Metaphysics

> Powerful cities stood there in very ancient times. Their walls have disappeared, but we have tablets on which the language of their inhabitants is engraved.
>
> Ashurbanipal, Assyrian King 650 BCE
> (referring to the Mesopotamian desert)

Ancient knowledge and artifacts tell a tale of the rise and fall of civilizations as regular as the seasons. To see beyond the conventions of current, established prejudices is to allow the soul to wander into grander vistas of other ages.

I affirm a natural ebb and flow of an ever-dying, ever-renewing cosmos and the unimaginable antiquity of humanity.

TRUTH IS WIDER THAN OUR NOTION OF IT...

Say not, "I have found the truth," but rather, "I have found a truth."

Kahlil Gibran

All great truths begin as blasphemies.

George Bernard Shaw

Knowledge is different is different states of consciousness...

Maharishi Mahesh Yogi

Those who know the truth are not equal to those who love it.

Confucius

Truth is more of a stranger than fiction.

Mark Twain

The truth is ungraspable and inexpressible. It neither is nor is not.

The Diamond Sutra

The opposite of a correct statement is a false statement. But the opposite of a profound truth may well be another profound truth.

Niels Bohr

> It is obvious that not everything that official science accepts at one time in history has to be true, and what one century affirms is in many cases denied by the next.
>
> Giorgio Livraga

> Truth is only hearsay until it becomes your own experience.
>
> Charles Lutes

What is considered as acceptable truth—, including some of the 'truths' of science—is more of matter of cultural convention. Life's paradoxes can defy our intellect's grasp of 'either /or truth.' Our intellectual formalisms pale before that great mystery called life, they stand at the portal only of pure being. I affirm that the poetic nature of the soul more aptly embraces the great truths of life and the understanding of being by its ability to make 'the flight of the alone to the Alone.'

For those of a more pragmatic bent of mind, who are self-assured of your assessment of reality, take a moment to ponder the sense data and your subsequent judgment thereupon. The next time you feel you are perfectly at rest and motionless, consider that your senses do not necessarily reveal the ways things really are. We are actually rotating, spinning and hurling through space. The astronomical facts reveal that the earth is rotating at a velocity of approx. 460 meters per second--or roughly 1,000 miles per hour and is revolving around our sun at 67,000 mph; all the while our solar system rides the grand rotation of our Milky Way galaxy at 514,000 mph. Simultaneously we're sweeping through the universe at 360 miles/sec or over 1 million mph in the direction of the constellation of Leo.

We have become adept as to the how of things, but cannot ascertain the why of things. We can calculate speeds to the nanosecond and distances to an angstrom, but have lost our dance step in the grand cosmic rhythm. We can no longer hear the 'music of the spheres.' We have become cosmic orphans, 'fugitives from the gods' in an age of the cybermachine.

We marvel at how some of our ancestors could arrive at marvelously accurate insights into the physics of the world, such as calculating pi and phi, the circumference of the earth, and atomic theory, all with the most subtle of instruments—the mind. And with that same instrument directed inward towards ever-deeper levels of silence, they intuited the origins of humanity and its journey towards the infinite. That was the age of poetry, as contrasted to the present, prosaic era. We live in a time that has no faith in legends, legends of our origin and destiny; we live in a time that has no faith... not the kind that would move mountains. Theirs, the ancients', was a grand remembrance of things as transmitted by the ancient myths and in their paeans to life, as in the Hymn to Dionysos—

> How they toss about in the immense universe as they whirl and seek each other, these myriad souls... and in the abyss mourn the forgotten homeland...

Myths are the distillations of deep truths, encoded in narrative form, from the ancient mystery schools, mankind's repositories of wisdom from time immemorial.

Beyond Gaia:

The Nature of the Cosmos

In the greater scheme of things, I affirm a 'living, breathing' universe that transcends our current notions of a material world made up of 'dead' matter. Plato described the universe as "one whole of wholes" and "a single living creature that encompasses all of the living creatures that are within it."

Expanding the boundaries of materialistic science along with the consciousness of the investigator or the knower may confirm the insights of the founders of western civilization, for those of a more daring spirit.

The **living nature of the cosmos** *is a natural corollary to the Gaia hypothesis, which asserts that our planet is a living, self-regulating organism.*

Consider the following **excerpt from: The Life of Apollonius of Tyana by Philostratus.** (Apollonius was a Neo-Pythagorean philosopher, circa 200 AD)

"Am I," said Apollonius, **"to regard the universe as a living creature?"**

"Yes," said the other, "if you have a sound knowledge of it, for it engenders all living things."

"Shall I then," said Apollonius, **"call the universe female, or of both the male and the opposite gender?"**

"Of both genders," said the other, "for by commerce with itself it fulfils the role both of mother and father in bringing forth living creatures; and it is possessed by a love for itself more intense than any separate being has for its fellow, a passion which knits it together into harmony." And it is not illogical to suppose that it cleaves unto itself; for as the movement of an animal is obtained by use of its hands and feet, and as there is a soul in it by which it is set in motion, so we must regard the parts of the universe also as adapting themselves through its inherent soul to all creatures which are brought forth or conceived.

For example, the sufferings so often caused by drought are visited on us in accordance with the soul of the universe, whenever justice has fallen into disrepute and is disowned by men; and this animal shepherds itself not with a single hand only, but with many mysterious ones, which it has at its disposal; *and though from its immense size it is controlled by no others yet it moves obediently to the rein and is easily guided."*

A metaphorical invention of the ancients, or insight into the reality of things?

You judge. Remember that "Truth is only hearsay until it becomes your own experience." It's much safer to retreat into accepted mental constructs than to think the "unthinkable," but we are at point in history that demands a "new" or a renewed way of "seeing" if we are to prosper as a species. And "seeing" implies going beyond mere thinking but arriving at a point of "stillness" that precedes all thinking. In ancient Greece this "stillness" was referred to as *hesychia*. "This term for "stillness" formed part of a precise technical vocabulary that was used to describe the state achieved "during deep meditation, ecstasies and dreams." *

This practice of "stillness" was much more important to the founding of our western philosophy and civilization than has been given credit in academic circles.

*Peter Kingsley—"Reality" / Ancient Philosophy, Mystery and Magic / "In the Dark Places of Wisdom"

The Myth of Materialism

(Going Down the Rabbit Hole of Physical Reality)

> "We must begin by discriminating between that which always is and never becomes; and that which is always becoming and never is"
>
> Plato - Timeaus Dialogue

What a novel thought for the modern mindset – discriminating between *"that which always is"* from *"that which never is."* Our educational system does not even teach us that there is a reality of non-change or Being, much less show us a way to experience it. Our daily lives are immersed and consumed with the sensory world of constant change. The knowledge of this immutable level of life is the lost wisdom that has been forgotten by modern civilization.

Plato leads the "uninitiated" into challenging our everyday notions of reality. We share the world of the senses with animal kingdom, but when an individual begins to discern *"that which always is"* or what is "real" from *"that which never is,"* or that which is always changing or has no permanence, this is the beginning of wisdom. Lifting our awareness to this level of life requires nothing less than *transcending* the senses and experiencing a "transcendental field of non-change."

Plato best illustrated this idea with his "Divided Line" which depicts an ascending order of realities from the gross through the subtle to the ultimate transcendental level of Being. Each level has it corresponding cognitive state of subjective awareness with the related world of objects. In other words, Plato is representing a scale of non-change and change or "Being and Becoming." For example, physical objects such as individual trees, people, and planets come and go, but the "Idea" or "Forms" of trees, people and planets are changeless, just like the rules of mathematics which are subtler and are more immutable than any physical objects. Cognitively speaking the corresponding levels go from mere belief and opinion in the "shadow world of sensory objects" which are constantly changing, then to knowledge and reasoning of universal principles until finally we arrive at direct intuition or cognition of the highest Universals – Transcendental "Goodness & Beauty" or the level of Being which is the source of all Becoming.

"Knowledge has three degrees--opinion, science, illumination. The means or instrument of the first is sense; of the second, dialectic; of the third, intuition"

Plotinus

In other words, one might say that the world of motion and things are a superimposition on the eternity of being and are less "real" by virtue of their transitoriness in contrast to eternity, somewhat like a wave on the ocean.

"Time is a moving image of eternity" – Plato

"The unreal has no being; the real never ceases to be. The final truth about them both has thus been perceived by the seers of ultimate Reality."

Vyasa

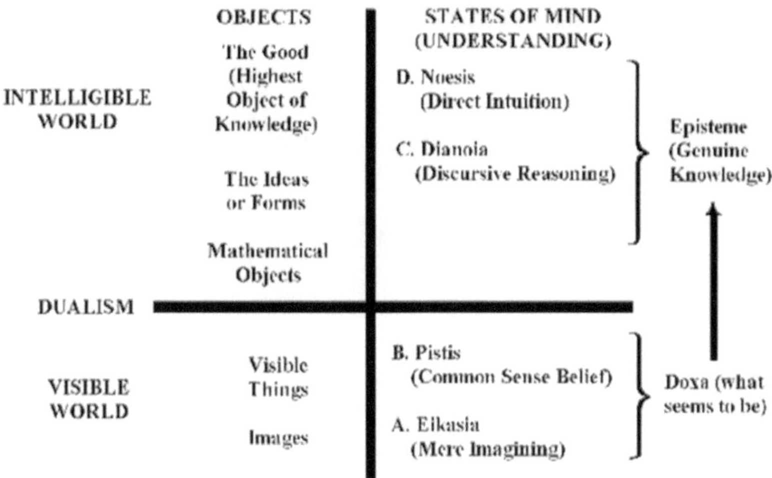

Plato also uses the Allegory of the Cave in connection with this same reasoning, that is, our view of reality is greatly lacking; the allegory is basically as follows: from birth we are like prisoners in a cave chained in such a way that we can only face a cave wall that has shadows cast on it from a source of light behind us and we know only those shadows as our reality. One of the prisoners escapes the cave with the help of someone (a teacher) to discover sunlight and the outside world of color and dimension. This is the "philosopher" who has discovered a more expansive reality. He comes back to inform the cave dwellers of this other reality but they spurn him and think him mad.

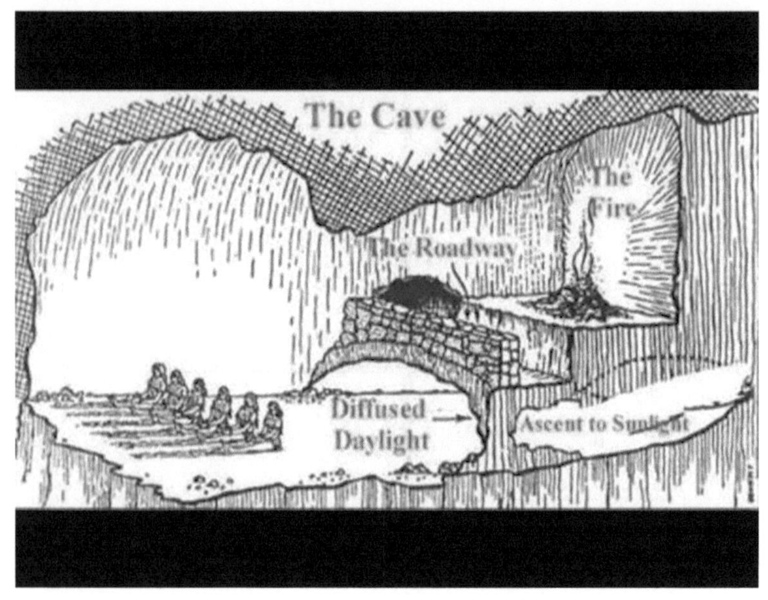

The allegory depicts the vast majority of the population who grow up with their awareness immersed in the phenomenal world all around them. We are virtually swimming and inundated with sensory data. We can be said to be living in a "sensory cave." Our everyday cognition or understanding of the world is equivalent to the dwellers in the cave that see only the shadowy world of images on the wall. The "shadow images" are the images and objects of the senses in which people mistaken these shadows for "Reality," with a capital "R." This is the cognitive level of opinion and belief. It is only when we can break free from the constraints of the "cave of the senses" to the realm of mind and beyond that we can start to have a greater awareness that transcends the world of becoming (change). As we move up the ladder of cognition we become aware of the intelligible world of number and ideas until we finally reach the immutable level of Being. In summary, this allegory refers to our expanding our awareness beyond the impermanent, material world to a permanent intelligible world which is the source of all change yet beyond it. It is a story about the human journey from darkness to light, from sleeping to waking, from ignorance to knowledge, from bondage to transcendence.

Beyond Plato's Cave

Science has operated on the premise or metaphysical assumption that matter is primary – scientific materialism.

> "While we may call it reality, external objective reality is not reality at all. An assumption that by its very nature cannot be verified is not a physical assumption, but is called a metaphysical assumption. (Such an assumption can also be called an axiom.) Thus, the bedrock of all science is not science at all but is metaphysics!"
>
> Stanley Sobottka
> Emeritus Professor of Physics
> University of Virginia

It is a given that every waking moment we are *confronted by our senses* (**"Plato's Cave"**) as to the pervasive reality of the material world. Quantum physics clearly demonstrates that the material world is a never-ceasing process of change or becoming; what we call material reality is in truth a constant state of creation and annihilation. Any given physical object is never the same moment to moment like running movie fames which appear to give continuity via discontinuous events. And the search for the building blocks of matter continues to take us down a "rabbit hole" where materiality "dissolves" into "interactions of forces" that ultimately point towards "no-thingness," as if the boundaries of objectivity seem to collapse at subtler, abstract levels of reality.

> "For as far as I know, there is no proof whatever of the existence of an objective reality apart from our senses, and I do not see why we should accept the outside world as such solely by virtue of our senses."
>
> M.C. Escher

> "Fools! They have no far-reaching thoughts who imagine that what was not before can come into being, or that anything can perish and be utterly destroyed."
>
> Empedocles

Such statements noted above seem like insane mutterings to the "practical, materialistic minded" yet even quantum physicists realize that the physical world seems very strange indeed at its more subtle levels. Consider the following statement which turns the "world on its head,"

"As a man who has devoted his whole life to the most clear headed science, to the study of matter, I can tell you as a result of my research about atoms this much: ***There is no matter as such.*** All matter originates and exists only by virtue of a force which brings the particle of an atom to vibration and holds this most minute solar system of the atom together. We must assume behind this force the existence of a conscious and intelligent mind. This mind is the matrix of all matter."

"I regard consciousness as fundamental. I regard matter as derivative from consciousness. We cannot get behind consciousness. Everything that we talk about, everything that we regard as existing, postulates consciousness."

Max Planck

"Whatever matter is, it is not made of matter."

Professor of Physics Hans-Peter Dürr Former Director of Max Planck Institute of Physics

"Physics is the study of the structure of consciousness. The "stuff" of the world is mindstuff."

Sir Arthur Eddington - British Physicist

These are astonishing statements given that they are coming from the "Father of Quantum Physics," and other notable physicists. Yet succeeding quantum scientists continue to acquiesce to their everyday sensory bias of materiality (**"Plato's Cave"**). The continued search for the ultimate constituent of matter seems to elude the piercing scientific intellect as it more and more realizes that particles (concepts) gave way to fields of energy (more subtle concepts) and fields give way to information (subtler concepts) which they will ultimately find emerge from the "black hole" or plenum of pure consciousness or in the words of Max Plank, "Mind – the Matrix of all Matter."

What are the characteristics of this plenum of pure consciousness? According to ancient Vedic wisdom the essential qualities of pure consciousness or this "matrix of all matter" are existence, intelligence and bliss. This perennial wisdom proclaims consciousness to be the primary reality and all else are but modifications of consciousness, just as the vast ocean is the one underlying reality to the transient individual waves of the sea. This realization at the level of hard science will herald nothing less than a 2nd Copernican Revolution! Just as our world view changed radically when we shifted from a geo-centric to a helio-centric one where the earth was no longer considered to be at the center of the universe, so too will be a radical shift from a materialistic-based to a consciousness-based reality.

This will be the most astonishing "discovery" of the 21st century that will open the flood gates to understanding ourselves and the powers of consciousness and the universes that proceed therefrom. How can we truly have a basis for knowledge without the knowledge of the knower? We must at last heed the injunction of the ancient temple:

"Know thyself and you will know the Universe and the Gods."

Science will at last realize that the shifting sands of matter are but the modifications of consciousness or a "field of intelligence," a supreme, cosmic "sleight of hand" of infinite consciousness and that the "Unified Field" of the objective universe is also the "Unified Field" of all subjectivity.

Modern science has operated from the metaphysical assumption that matter is primary without question as a self-evident truth. To be sure, the metaphysical assumptions of science are guiding principles only, a working model that has been showing more and more cracks. Many so-called anomalies of psi phenomena or mind/machine interaction and other mind/matter conundrums including "afterlife" experiences will fall into place as we realize that matter is the "coalescence" of universal consciousness.

Finally, the subjective spiritual disciplines will find logical congruence with the objective material disciplines in the common ground of Consciousness as the unifying ground of existence.

The self-replicating evidence of this fact must be found in the consciousness of the knower/scientist by firstly cognizing one's essential nature which is at once the ground state of one's own being and that of the universe.

"The One is all things and yet no one of them. It is the source of all things, not itself all things, but their transcendent Principle"

Plotinus

The requirement of the new science will be nothing less than Self-Knowledge, *"Know that by which all else is known" the ground state or "the unified field."* It must be our own direct experience or living reality otherwise it's a theory only – "knowledge in the books stays in the books;" then and only then will the "cosmic scientist" realize that the ground of his being and that of the universe is the same unchanging transcendental reality.

"Consciousness is the common denominator underlying the possibility of any philosophy, world view, religious attitude, art, or science. I, therefore, affirm the systematic primacy of consciousness as such."

Franklin Merrell-Wolff (philosopher/mathematician)

"Wave-particle duality, the uncertainty principle, the collapse of the wave function, and entanglement all point to awareness being an intrinsic aspect of reality. Yet we are still trying to understand these phenomena within a worldview that assumes the real world to be that of space, time, and matter, and relegates consciousness to some artifact of brain processes. Yet the one thing of which we are certain is that we are aware. And it is the one thing the current worldview cannot account for. This profound anomaly will ultimately lead to the full paradigm shift to which contemporary physics is, unwittingly, pointing. With consciousness as primary, everything remains the same and everything changes. Mathematics, physics, biology, chemistry are unchanged. What changes is our assumption as to what they are describing. They are not describing the unfolding of a physical world, but the unfolding of a universal self-aware field. We are led to the conclusion that the entire cosmos is a vast field of knowing, knowing itself, and in that knowing creating for itself the appearance of a material world.

https://www.reddit.com/r/badphilosophy/comments/3rr9yg/with_consciousness_as_primary_everything_remains/

Ultimately, all theories, all reasoning and all observations are made in consciousness. We have assumed a physical world of objects by virtue of our senses only and that it exists outside the domain of consciousness. Ironically, our probing with the most exacting science finds the more abstract levels of this "outside reality" at the very doorsteps of consciousness.

An excellent starting reference point is Peter Russell's "The Primacy of Consciousness" documentary.

http://topdocumentaryfilms.com/the-primacy-of-consciousness/

Transcendence in an Age of Materialism

> The world is too much with us; late and soon, Getting and spending, we lay waste our powers;— Little we see in Nature that is ours; We have given our hearts away, a sordid boon! This Sea that bares her bosom to the moon; The winds that will be howling at all hours, And are up-gathered now like sleeping flowers; For this, for everything, we are out of tune; It moves us not. Great God! I'd rather be A Pagan suckled in a creed outworn; So might I, standing on this pleasant lea, Have glimpses that would make me less forlorn; Have sight of Proteus rising from the sea; Or hear old Triton blow his wreathèd horn.
>
> The World Is Too Much With Us
>
> BY WILLIAM WORDSWORTH

Our current culture's focus on materialism with its promise of a life more abundant is but one of the many counterfeits of transcendence or false promises for fulfilling the human spirit in its quest for enduring happiness.

The Orphics* of the ancient Greek world had a saying, Soma Sema, which translates as 'the body is a tomb' (to the spirit). Our present-day ethos has become a straightjacket or a means of numbing our finer sensibilities and our abilities to be attuned to nature's delicate rhythms.

---

* Orphics—members of an early Greek mystery religion

> Let us be silent that we may hear the whispers of the gods
>
> Emerson

Today's postmodern culture presents a view of the world sanitized to the point of banality. It promotes a status quo of the self and not a true discovery of our deepest selves beyond the transient ego. We are peddled palliative commodities and hectic amusements to ease the pain in a world of violence. Our techno-culture seems incapable of addressing the deeper questions of life—"whence, whither, and why?"—in any meaningful and satisfying way to "our deeper longing." We are veritably lost in a cultural labyrinth devoid of any meaningful mythos. We are living in an age of prolific information, yet we are virtually drowning in information, in an age of '*informed bewilderment*.'

> The 21st century will not be a dark age, neither will it deliver to most people the bounties promised by the most extraordinarily technological revolution in history. Rather, it may well be characterized by informed bewilderment.
>
> Manuel Castells

On the other hand, the myths of ancient cultures provided a roadmap or perspective on life, as pointed out by Joseph Campbell: "*Myths are clues to the spiritual potentialities of the human life.*" We now live in the age of the anti-hero, where sensationalism provides stimulation devoid of nutrition to our deeper selves. Today's hero is a 'Rambo' thrashing about in convulsive, brutish fits of violence, incapable of being informed by sublime sources.

How and why have we arrived at such a critical state of affairs, to find ourselves in a spiritual wasteland?

> I have discovered that all human evil comes from this, man's being unable to sit still in a room.
>
> Blaise Pascal (philosopher and mathematician)

By our not transcending our everyday level of thinking or knowing how to go to the source of our own being!

> Because what the soul seeks is the One and it would look upon the source of all reality... Rather must it withdraw from sense objects... It must rise to the principle possessed within itself.
>
> Plotinus (philosopher)

This 'One' has in modern terms been identified with the "unified field of consciousness," the origin of all subjective and objective existence, the source of our being: pure, unadulterated intelligence, existence, energy, and bliss. This is not a matter of scientific speculation or philosophical or theological discourse, but a matter of direct experience.

The human physiology is designed for this refined experience, and it is the lack of our experiencing and actualizing this level of life that results in suffering and violence in life.This is accomplished through the discipline of meditation, of directing our attention to deeper levels of silence within.

One of the clearest descriptions of transcending in our western culture was by St. Augustine of Hippo:

The power of reason, realizing that in me, too, it was liable to change, led me to consider the source of its own understanding. It withdrew my thoughts from their natural course... so that it might discover what light it was that had been shed upon it when it proclaimed for certain that what was immutable was better than that which was not, and how it had come to know the immutable itself. For unless by some means, it had known the immutable, it could not possibly have been certain that it was preferable to the mutable. And so in an instant of awe, my mind arrived at *that which is.*

He transcended or went beyond everyday thinking and the world of change to realize that the source of reason or the mind is a level of non-change or absolute being. He further stated had he not known this level of non-change, he would not know that it was preferable to the world of change. This state or level of non-change is a level of unity. It is the "unified field of life" as described most clearly, since time immemorial, in the ancient Vedic tradition—the oldest known tradition of knowledge on the planet, which admonished us to "Know that by which all else is known." And it is very likely that this tradition was the germinating seed of western culture.

What constitutes the anatomy of our present situation? In considering our social condition we can look to the structure of our brain to provide a clue as to the level of our society's functioning. Our brains are composed of three distinct sections or separate "brains." Our triune brains consist of a 'reptilian brain,' 'early mammalian brain,' and 'neo-cortical brain'—or the instinctive, emotional, and reasoning centers, respectively. Today's media seems to target the 'reptilian' or survival instincts along with the emotional or 'early mammalian' brain centers, whereas the rudimentary neo-cortical center for discursive thinking seems bypassed with glib pronouncements or sound bites. Our current educational system is clearly defective inasmuch as it provides a fragmented, information-based level of learning and does not develop a whole-brain or integrated, functioning individual.This is clearly self-evident from the enduring problems of our society.

Our present consumer-mentality society appeals directly to our survival instincts and creature comforts. When the 'reptilian brain' dominates, competition for the sake of power and domination are the hallmarks. Our 'emotional brain' is best characterized by our instincts to protect the domestic hearth or family. Our neo-cortical functioning has given rise to systems of philosophy, science, and thinking about and communicating ideas.

An optimal, whole-brain functioning or "enlightened" individual is one whose thoughts and actions are in constant synchrony with the common good, i.e. environment, other individuals, and society. Synchrony of brain functioning correlates with the synchrony of functioning with natural law or the universe. This is the goal that we should ultimately be seeking. The ancient Greeks had a word for describing this wholeness of character: sophrosyne, a term which suggests a kind of order or balance that stretches from the innermost reaches of the soul to the surface of the body—the whole person in synch with the universe!

We have been from suffering a cosmic amnesia, but fortunately for us the ancient science (of consciousness) that has been the wellspring of civilizations—though periodically forgotten—has been resuscitated at this crucial time in human history. It is to be found most clearly in the Vedic tradition mentioned earlier.

Modern or materialistic science operates from an "after-the-fact" or a posteriori, trial and error, method. Conversely, this ancient science is based on an a priori, or "before-the-fact," "science of first causes," or "knowledge of the absolute" as opposed to an "approximate knowledge of the relative" type of science. The testimony of this profound understanding of natural law lies all around the globe in the fantastic ruins of ancient cultures. The riddles of ancient fantastic feats will not be solved through materialist consciousness.

> The significant problems we face can never be solved at the level of thinking that created them.
>
> Albert Einstein

Whence and why comes human suffering? Forgetting to transcend or not connecting to our divine source seems a likely candidate, does it not? Or in the words of a contemporary scientist:

> ...if we deny ourselves transcending, we experience suffering because we do not use our full mental potential
>
> Demetri Kanellakos

The need for transcendence has never been more imperative than it is today. Our current situation is like the Sorcerer's Apprentice—we have unleashed the 'magic' or technology of science without the wisdom to control it or ourselves. "*What has been will be again, what has been done will be done again, there is nothing new under the sun.*" Worldwide legends tell of other civilizations that have risen and perished when cleverness exceeded wisdom, egotism exceeded spirituality. But in our collective amnesia we assert that our current level of knowledge is the sole province of our present times.

Current 'academic wisdom' discounts the testimony of ancient cultures' claim to remote antiquity, despite being confronted and confounded with numerous and increasing out-of-place artifacts that continue to defy conventionally accepted beliefs. A modicum of research will reveal that the facts do not fit neatly and in many instances defy the accepted, mainstream paradigm of a linear model of human history. A seminal work, Hamlet's Mill, sheds light on precise astronomical knowledge encoded in ancient myths revealing a knowledge of the "Great Year"—a cosmic clock by which ancients could ascertain the rise and fall of civilizations. By the ancients' own reckoning, there exist "cosmic seasons" that give rise to the flowering and decay of civilizations and human understanding. This time scale is also known as the Precession of the Equinoxes, which is approximately 25,000 years in duration. And this Great Year is only 1/180th of a larger cycle, which is the 'tip of the iceberg' of even larger and larger cosmic cycles described by the ancients.

According to the ancient Vedic tradition, when knowledge and personal awareness of all 'three worlds' is lively—i.e. the gross, subtle, and transcendent aspects—this constitutes a 'Golden Age' or one of an enlightened society. This is a time that is lived effortlessly in harmony with all the laws of nature or the 'will of God' or the paradise of legend; it is what is meant when it was said that 'man walked with God.'

> And 'tis told by ancient sages, during Rama's happy reign, Death untimely, dire diseases came not to his subject men,
>
> Widows wept not in their sorrow for their lords untimely lost, Mothers wailed not in their anguish for their babes by death's crost,
>
> Robbers, cheats, and gay deceivers tempted not with lying word, Neighbor loved his righteous neighbor and the people loved their lord!
>
> Trees their ample produce yielded as returning seasons went, And the earth in grateful gladness never failing harvest lent,
>
> Rains descended in their season, never came the blighting gale, Rich in crop and rich in pasture was each soft and smiling vale,
>
> Loom and anvil gave their produce and the tilled and fertile soil, And the nation lived rejoicing in their old ancestral.

When knowledge of the transcendent aspect of life is lost and only the appreciation of the subtle and gross remain, this is characteristic of an age of a "subtle science" or what has been called magic, whereby knowledge of the subtle mechanics of nature is accessible, but is wielded on a more limited basis and can be used for more selfish purposes. This state of affairs was evident in the twilight of the Egyptian and Mesoamerican cultures. This state eventually gives way to superstition or fear of the unknown as even the knowledge of the subtle mechanics of life starts to get lost.

> Atheism is reason deluded, superstition an emotion growing out of deluded reasoning.
>
> Plutarch

Finally we come to the materialistic age, which springs up where superstition leaves off, as keener minds try to make sense only of the phenomenal or grosser fields of life. Empirical science starts from this basis and proceeds in a reverse course, as it were, by trial and error, to discover nature's more rarified principles, but from an outside-looking-in perspective—i.e. outside that field of pure consciousness.

So we stand at a unique moment in history where the ancient truths that were guarded in the inner sanctorum of the temples are now accessible to all who have eyes to see and ears to hear.

> Let the specimen suffice to those who have ears.
>
> St. Clement of Alexandria

> There is nothing concealed that will not be disclosed, or hidden that will not be made known.
>
> Luke 12.2

As the inscription on the temple of Delphi read: "Know thyself and you will know the universe and the gods."

The need of the times is upon us to connect with our deepest self, to listen to the whispers of the soul or the inner silence, and to be a conduit of light for the world.

Let us contact and spread the bliss that lies within each of us.

In the Greater Scheme of Things... these worlds are fundamentally a "play of consciousness," and consciousness or the unified field is the "fountain of all material manifestation"—this was the secret of the ancients.

> Take me from the unreal to the real
>
> Take me from darkness to the light
>
> Take me from death to immortality.

Epilogue

The Unity of Life

"The total number of minds in the universe is one. In fact, consciousness is a singularity phasing within all beings."- Erwin Schrodinger (Quantum Physicist)

"He whose self is established in Yoga, whose vision everywhere is even, sees the Self in all beings and all beings in the Self"

Bhagavad Gita Ch.VI v 29

"All fear is born of duality" - Brihadaranyaka Upanishad

The day to day struggles and the struggles of history can be for the most part relegated to the culprit ego with its notion of "mine" and "thine" beginning with the first deadly vice, vanity or pride and the ensuing other deadly vices of covetousness, greed, envy and so forth.

In today's contentious society I am reminded of the Victorian saying, "It takes two fools to argue' - two wise men will not argue, one wise man will not argue with a fool, therefore it takes two fools to argue; which if true, then we have become a society of fools with all the unending contention seen all around us. It is not unreasonable to assume that if we are to survive as a species it is imperative that we must change. I submit that the change needed is no less than a radical new way seeing and interacting with the world. The world awaits a "phase transition," a new "state of consciousness" – one that transcends our anthropocentric ego. We must learn to see beyond our own "skins" or small selves and encompass our "vast Self" of which we are all part of. In religious parlance, this is often referred to as Spirit, an oftentimes nebulous term. Our contentions arise from the predominant sense of ego and separateness, "mine" and "thine." We must rise above this duality and discover the deeper unity within us all.

> "Selfish egoism is the root of the wide extending branches of misery in the forest of this world."
>
> Vasistha

For all our technological know-how, we seem to be none the wiser and only more confused as to our place in the universe as we seem to be the victims of our "informed bewilderment." Ours is a spiritually rootless technocratic society based on materialism as its core assumption.

> "Where is the knowledge which is lost in information? And where is the wisdom that is lost in knowledge?"
>
> T.S. Eliot

"We shall not cease from exploration
And the end of all our exploring
Will be to arrive where we started
And know the place for the first time."
T.S. Eliot, Four Quartets

Where is the point of departure and arrival in all our exploration – is it not us, the perceiver, our own consciousness, our own sense of existence?

"We are by nature observers, and thereby learners. That is our permanent state." - Ralph Waldo Emerson

How can we claim to know anything when we don't know who we truly are? Self-knowledge is vastly more than belief or glib pronouncements of the contents of our thinking and feeling, or formalized theories, but an arrival at the source of our own consciousness which is the source of all life – it is the Great Silence or "Unified Field" from which all activity manifests. This is the source of all subjective and objective existence.

"What is this thing I call self? You mean you understood everything else in the world and you didn't understand this? You mean you understood astronomy and black holes and quasars and you picked up computer science, and you don't know who you are? My, you are still asleep. You are a sleeping scientist."

Anthony De Mello

We subdivide life and the world around us in our thinking. We live in a "shadow world" of sensory perception, a kind of "sensory cave" which is but the outer rim of more profound roots of life which are the very roots of our being. Ancient wisdom states that to become conscious of the deepest roots of life is to awaken to the unity of life.

"He who sees all beings in the Self, and the Self in all beings, hates none. To the illumined soul, the Self is all. For him who sees everywhere oneness, how can there be delusion or grief?

The Isha Upanishad

We are not just our brother's keeper, but in a most profound sense, we <u>are</u> our brother at the very heart of life! There is transcendence beyond the ego-sense and "the other" which is an awakening to more profound levels of reality.

The warp and weft of life is a "unity of Being," the matrix of all subjective and objective fields of life, a "Unified Field," whereas the multiplicity of appearances is but a dance on this limitless canvas of Being.

The following fictional account by Leo Tolstoy best illustrates the insights and "mystery knowledge" of ancient arcane wisdom about the unity of life and the essential brotherhood of humanity. This knowledge was not theoretical, but the direct experience of the initiates that had awoken beyond the din of the senses onto a transcendental level of Being. Plato's Allegory of the Cave was an allusion to the restricted field of perception of the common man, asleep in the virtual "cave of sensory life" unaware of the higher dimensionality of Life, a prisoner of his "ego sense." We are part and parcel of the Unbounded Ocean of Life – Pure Consciousness! Who we really are transcends the boundaries of the senses.

Read and absorb yourself, in this fictional account, as King Esarhaddon who "drunk with power" escapes the prison of his own ego even momentarily to discover a more transcendental vista.

Esarhaddon, King of Assyria

THE Assyrian King, Esarhaddon, had conquered the kingdom of King Lailie, had destroyed and burnt the towns, taken all the inhabitants captive to his own country, slaughtered the warriors, beheaded some chieftains and impaled or flayed others, and had confined King Lailie himself in a cage.

As he lay on his bed one night, King Esarhaddon was thinking how he should execute Lailie, when suddenly he heard a rustling near his bed, and opening his eyes saw an old man with a long grey bead and mild eyes.

'You wish to execute Lailie?' asked the old man.

'Yes,' answered the King. 'But I cannot make up my mind how to do it.'

'But you are Lailie,' said the old man.

'That's not true,' replied the King. 'Lailie is Lailie, and I am I.'

'You and Lailie are one,' said the old man. 'You only imagine you are not Lailie, and that Lailie is not you.'

'What do you mean by that?' said the King. 'Here am I, lying on a soft bed; around me are obedient men-slaves and women-slaves, and to-morrow I shall feast with my friends as I did to-day; whereas Lailie is sitting like a bird in a cage, and to-morrow he will be impaled, and with his tongue hanging out will struggle till he dies, and his body will be torn in pieces by dogs.'

'You cannot destroy his life,' said the old man.

'And how about the fourteen thousand warriors I killed, with whose bodies I built a mound?' said the King. 'I am alive, but they no longer exist. Does not that prove that I can destroy life?'

'How do you know they no longer exist?'

'Because I no longer see them. And, above all, they were tormented, but I was not. It was ill for them, but well for me.'

'That, also, only seems so to you. You tortured yourself, but not them.'

'I do not understand,' said the King.

'Do you wish to understand?'

'Yes, I do.'

'Then come here,' said the old man, pointing to a large font full of water.

The King rose and approached the font.

'Strip, and enter the font.'

Esarhaddon did as the old man bade him.

'As soon as I begin to pour this water over you,' said the old man, filling a pitcher with the water, 'dip down your head.'

The old man tilted the pitcher over the King's head and the King bent his head till it was under water.

And as soon as King Esarhaddon was under the water he felt that he was no longer Esarhaddon, but someone else. And, feeling himself to be that other man, he saw himself lying on a rich bed, beside a beautiful woman. He had never seen her before, but he knew she was his wife. The woman raised herself and said to him:

'Dear husband, Lailie! You were wearied by yesterday's work and have slept longer than usual, and I have guarded your rest, and have not roused you. But now the Princes await you in the Great Hall. Dress and go out to them.'

And Esarhaddon -- understanding from these words that he was Lailie, and not feeling at all surprised at this, but only wondering that he did not know it before -- rose, dressed, and went into the Great Hall where the Princes awaited him.

The Princes greeted Lailie, their King, bowing to the ground, and then they rose, and at his word sat down before him; and the eldest of the Princes began to speak, saying that it was impossible longer to endure the insults of the wicked King Esarhaddon, and that they must make war on him. But Lailie disagreed, and gave orders that envoys shall be sent to remonstrate with King Esarhaddon; and he dismissed the Princes from the audience. Afterwards he appointed men of note to act as ambassadors, and impressed on them what they were to say to King Esarhaddon. Having finished this business, Esarhaddon -- feeling himself to be Lailie -- rode out to hunt wild asses. The hunt was successful. He killed two wild asses himself, and having returned home, feasted with his friends, and witnessed a dance of slave girls. The next day he went to the Court, where he was awaited by petitioners suitors, and prisoners brought for trial; and there as usual he decided the cases submitted to him. Having finished this business, he again rode out to his favorite amusement: the hunt. And again he was successful: this time killing with his own hand an old lioness, and capturing her two cubs. After the hunt he again feasted with his friends, and was entertained with music and dances, and the night he spent with the wife whom he loved.

So, dividing his time between kingly duties and pleasures, he lived for days and weeks, awaiting the return of the ambassadors he had sent to that King Esarhaddon who used to be himself. Not till a month had passed did the ambassadors return, and they returned with their noses and ears cut off.

King Esarhaddon had ordered them to tell Lailie that what had been done to them -- the ambassadors -- would be done to King Lailie himself also, unless he sent immediately a tribute of silver, gold, and cypress-wood, and came himself to pay homage to King Esarhaddon.

Lailie, formerly Esarhaddon, again assembled the Princes, and took counsel with them as to what he should do. They all with one accord said that war must be made against Esarhaddon, without waiting for him to attack them. The King agreed; and taking his place at the head of the army, started on the campaign. The campaign lasts seven days. Each day the King rode round the army to rouse the courage of his warriors. On the eighth day his army met that of Esarhaddon in a broad valley through which a river flowed. Lailie's army fought bravely, but Lailie, formerly Esarhaddon, saw the enemy swarming down from the mountains like ants, over-running the valley and overwhelming his army; and, in his chariot, he flung himself into the midst of the battle, hewing and felling the enemy. But the warriors of Lailie were but as hundreds, while those of Esarhaddon were as thousands; and Lailie felt himself wounded and taken prisoner. Nine days he journeyed with other captives, bound, and guarded by the warriors of Esarhaddon. On the tenth day he reached Nineveh, and was placed in a cage. Lailie suffered not so much from hunger and from his wound as from shame and impotent rage. He felt how powerless he was to avenge himself on his enemy for all he was suffering. All he could do was to deprive his enemies of the pleasure of seeing his sufferings; and he firmly resolved to endure courageously without a murmur, all they could do to him. For twenty days he sat in his cage, awaiting execution. He saw his relatives and friends led out to death; he heard the groans of those who were executed: some had their hands and feet cut off, others were flayed alive, but he showed neither disquietude, nor pity, nor fear. He saw the wife he loved, bound, and led by two black eunuchs. He knew she was being taken as a slave to Esarhaddon. That, too, he bore without a murmur. But one of the guards placed to watch him said, 'I pity you, Lailie; you were a king, but what are you now?' And hearing these words, Lailie remembered all he had lost. He clutched the bars of his cage, and, wishing to kill himself, beat his head against them. But he had not the strength to do so and, groaning in despair, he fell upon the floor of his cage.

At last two executioners opened his cage door, and having strapped his arms tight behind him, led him to the place of execution, which was soaked with blood. Lailie saw a sharp stake dripping with blood, from which the corpse of one of his friends had just been torn, and he understood that this had been done that the stake might serve for his own execution. They stripped Lailie of his clothes. He was startled at the leanness of his once strong, handsome body. The two executioners seized that body by its lean thighs; they lifted him up and were about to let him fall upon the stake.

'This is death, destruction!' thought Lailie, and, forgetful of his resolve to remain bravely calm to the end, he sobbed and prayed for mercy. But no one listened to him.

'But this cannot be,' thought he. 'Surely I am asleep. It is a dream.' And he made an effort to rouse himself, and did indeed awake, to find himself neither Esarhaddon nor Lailie -- but some kind of an animal. He was astonished that he was an animal, and astonished, also, at not having known this before.

He was grazing in a valley, tearing the tender grass with his teeth, and brushing away flies with his long tail. Around him was frolicking a long-legged, dark-gray ass-colt, striped down its back. Kicking up its hind legs, the colt galloped full speed to Esarhaddon, and poking him under the stomach with its smooth little muzzle, searched for the teat, and, finding it, quieted down, swallowing regularly. Esarhaddon understood that he was a she-ass, the colt's mother, and this neither surprised nor grieved him, but rather gave him pleasure. He experienced a glad feeling of simultaneous life in himself and in his offspring.

But suddenly something flew near with a whistling sound and hit him in the side, and with its sharp point entered his skin and flesh. Feeling a burning pain, Esarhaddon -- who was at the same time the ass -- tore the udder from the colt's teeth, and laying back his ears galloped to the herd from which he had strayed. The colt kept up with him, galloping by his side. They had already nearly reached the herd, which had started off, when another arrow in full flight struck the colt's neck. It pierced the skin and quivered in its flesh. The colt sobbed piteously and fell upon its knees. Esarhaddon could not abandon it, and remained standing over it. The colt rose, tottered on its long, thin legs, and again fell. A fearful two-legged being -- a man -- ran up and cut its throat.

'This cannot be; it is still a dream! thought Esarhaddon, and made a last effort to awake. 'Surely I am not Lailie, nor the ass, but Esarhaddon!'

He cried out, and at the same instant lifted his head out of the font. . . . The old man was standing by him, pouring over his head the last drops from the pitcher.

'Oh, how terribly I have suffered! And for how long!' said Esarhaddon.

'Long?' replied the old man, 'you have only dipped your head under water and lifted it again; see, the water is not yet all out of the pitcher. Do you now understand?'

Esarhaddon did not reply, but only looked at the old man with terror.

'Do you now understand,' continued the old man, 'that Lailie is you, and the warriors you put to death were you also? And not the warriors only, but the animals which you slew when hunting and ate at your feasts were also you. You thought life dwelt in you alone but I have drawn aside the veil of delusion, and have let you see that by doing evil to others you have done it to yourself also. Life is one in them all, and yours is but a portion of this same common life. And only in that one part of life that is yours, can you make life better or worse -- increasing or decreasing it. You can only improve life in yourself by destroying the barriers that divide your life from that of others, and by considering others as yourself, and loving them. By so doing you increase your share of life. You injure your life when you think of it as the only life, and try to add to its welfare at the expense of other lives. By so doing you only lessen it. To destroy the life that dwells in others is beyond your power. The life of those you have slain has vanished from your eyes, but is not destroyed. You thought to lengthen your own life and to shorten theirs, but you cannot do this. Life knows neither time nor space. The life of a moment, and the life of a thousand years: your life and the life of all the visible and invisible beings in the world, are equal. To destroy life, or to alter it, is impossible; for life is the one thing that exists. All else, but seems to us to be.'

Having said this the old man vanished.

Next morning King Esarhaddon gave orders that Lailie and all the prisoners should be set at liberty and that the executions should cease.

On the third day he called his son Assurbanipal, and gave the kingdom over into his hands; and he himself went into the desert to think over all he had learnt. Afterwards he went about as a wanderer through the towns and villages, preaching to the people that all life is one, and that when men wish to harm others, they really do evil to themselves.

In this account it's as if King Esarhaddon underwent an "initiation" whereby his consciousness was lifted beyond "the prison of the senses" to experience the realm of "Spirit" or "transcendental field of being" which is infinitely correlated with all Life. This was undoubtedly one of the "mysteries" that was revealed to the "spiritually initiated" in the perennial wisdom schools of the ancient temples. Let's examine more closely the climatic insight of this story.

'Do you now understand.., 'that Lailie is you, and the warriors you put to death were you also?

Our essential nature is not the ego which is a temporary mask of this transient existence which we blithely mistaken for our "essential self." Ancient wisdom proclaims that who we truly are is of a more enduring nature or Cosmic Self which is the ground of existence of us all.

How can the Self be killed or kill (* Bhagavad Gita) When there is only One?

You thought life dwelt in you alone but I have drawn aside the veil of delusion, and have let you see that by doing evil to others you have done it to yourself also. Life is one in them all, and yours is but a portion of this same common life.

Life transcends biological identity nor is it the result of chemical interactions, rather biochemical processes are part of the infinite display of Infinite Intelligence. We are all part of this universal field of being."

Bodies die, not the Self that dwells therein. (* Bhagavad Gita)
Know the Self to be beyond change and death.
Therefore strive to realize this Self.

And only in that one part of life that is yours, can you make life better or worse -- increasing or decreasing it. You can only improve life in yourself by destroying the barriers that divide your life from that of others...

Our consciousness is sovereign; we can increase our awareness and thereby "expand our domain of influence" through conscious awareness of this infinite field of Life or we can maintain the barriers of nescience.

Some there are who have realized the Self (* Bhagavad Gita)
In all its wonder. Others can speak of it
As wonderful. But there are many
Who don't understand even when they hear.

...and by considering others as yourself, and loving them. By so doing you increase your share of life.

By consciously aligning with this Infinite Field of Life, we consciously partake of this cosmic status. We are Cosmic Intelligence playing hide and seek in different bodies. "Mine and thine" are assumptions derived from the sensory level only, a mistake of the intellect.

The wise, who live free from pleasure and pain, (* Bhagavad Gita)
Are worthy of immortality.

You injure your life when you think of it as the only life, and try to add to its welfare at the expense of other lives. By so doing you only lessen it.

Life is a "universal field" which is infinitely correlated with all its manifestations, no one exists in isolation; "No man is an island." We are all part and parcel intimately correlated in the web of Life. This reality is the basis of the "Golden Rule." Herein is the basis of the reciprocity of all life.

Those who look upon the Self as slayer (* Bhagavad Gita)
Or as slain have not realized the Self.

To destroy the life that dwells in others is beyond your power. The life of those you have slain has vanished from your eyes, but is not destroyed

Life like the law of conservation of energy can neither be created nor destroyed but is either Manifest or Unmanifest (to the senses) in the "eternal sea of Being."

Even as we cast off worn-out garments (* Bhagavad Gita)
And put on new ones, so casts off the Self
A worn-out body and enters into
Another that is new.

Not pierced by arrows nor burnt by fire, (* Bhagavad Gita)
Affected by neither water nor wind,
The Self is not a physical creature.

You thought to lengthen your own life and to shorten theirs, but you cannot do this. Life knows neither time nor space. The life of a moment, and the life of a thousand years: your life and the life of all the visible and invisible beings in the world, are equal. To destroy life, or to alter it, is impossible; for life is the one thing that exists. All else, but seems to us to be.

Life is that universal field of being which is pure boundless "oceanic existence," transcendental, non-changing which gives rise to all change as the ocean gives rise to the individual waves of the sea.

Who knows the Self to be birthless, deathless, (*
Bhagavad Gita)
Not subject to the tyranny of time,
How can the Self slay or cause to be slain?
Deathless is the Self in every creature.
Know this truth, and leave all sorrow behind.

This insight and wisdom of the unity of life was undoubtedly an integral aspect of the ancient wisdom that was realized directly for those that journeyed the path of self-realization. I would submit that initiation into the Lesser Mysteries of the arcane wisdom of the Mystery Schools was the realization that the individual is an undying entity, immortal in nature, whereas the initiation into the Greater Mysteries was the direct realization of the Unity of Life!

Some of these fortunate individuals' insights have echoed down the corridors of time in answer to Life's riddle.

"What is real never ceases to be.
The unreal never is. The sages
Who realize the Self know the secret
Of what is and what is not."

Bhagavad Gita

“Vincit Qui Se Vincit”

He Conquers Who Conquers Himself

- Publilius Syrus – 1st Century BC

www.ingramcontent.com/pod-product-compliance
Ingram Content Group UK Ltd.
Pitfield, Milton Keynes, MK11 3LW, UK
UKHW041936190726
13854UKWH00004B/1627